Real Science-4-Kids

21 Super Simple Astronomy Experiments

Rebecca W. Keller, Ph.D.

Illustrations: Rebecca W. Keller, Ph.D.
Photographs: Courtesy of NASA—nasaimages.org

Copyright © 2011 Gravitas Publications, Inc.

All rights reserved. No part of this publication may be reproduced, stored in a retrieval system, or transmitted, in any form or by any means, electronic, mechanical, photocopying, recording, or otherwise, without prior written permission from the publisher. However, this publication may be photocopied without permission from the publisher only if the copies are to be used for teaching purposes within a family.

Real Science-4-Kids: 21 Super Simple Astronomy Experiments

ISBN 10: 1936114208
ISBN 13: 9781936114207

Published by Gravitas Publications, Inc.
4116 Jackie Road SE, Suite 101
Rio Rancho, NM 87124
www.gravitaspublications.com

Printed in United States

What are Super Simple Science Experiments?

Super Simple Science Experiments are experiments that focus on one aspect of scientific investigation. Doing science requires students to develop different types of skills. These skills include the ability to make good observations, turning observations into questions and/or hypotheses, building and using models, analyzing data, using controls, and using different science tools including computers.

Super Simple Science Experiments break down the steps of scientific investigation so that students can focus on one aspect of scientific inquiry. The experiments are simple and easy to do, yet they are *real* science experiments that help students develop the skills they need for *real* scientific investigations.

Each experiment is one page and lists a short objective, the materials needed, a brief outline of the experiment, and includes any graphics or illustrations needed for the experiment. The skill being explored is listed in the upper right hand corner of each page. Any additional pages required are included at the back of the book.

Getting Started

Below is a list of the materials for all the astronomy experiments in this book. You can collect all the materials ahead of time and place them in a storage bin or drawer.

Materials at a Glance	
Super Simple Science Experiments Laboratory Notebook	water-based craft paint in the following colors:
compass	white
pencil	gray
ping-pong ball	yellow
softball	red
flashlight	orange
balloons	blue
cardboard	green
scissors	styrofoam balls with the following dimensions
popsicle sticks	1 inch
tape	1.5 inches
protractor	2 inches (3 balls)
toothpicks	4 inches (2 balls)
glue	6 inches
marker	8 inches
	12 inches

Table of Contents

Title   Page

1. Finding the North Star   1
2. Finding the Big Dipper   2
3. Finding the Little Dipper   3
4. Finding the Dragon   4
5. Finding Cassiopeia   5
6. Moon Phase Calendar   6
7. Solar Eclipse   7
8. Lunar Eclipse   8
9. Ocean Waves and the Moon   9
10. Building a Horizontal Sundial   10
11. Model of the Moon   11
12. Model of the Sun   12
13. Model of Mars   13
14. Model of Venus   14
15. Model of Jupiter   15
16. Model of Saturn   16
17. Model of Neptune and Uranus   17
18. Model of Mercury   18
19. Model of Earth   19
20. Model of the Solar System   20
21. Finding the Center of the Milky Way Galaxy   21
Sundial Diagram   22
Calendar template   23

1. Finding the North Star

observation

Objective

To locate the North Star (Polaris) using a compass. (Visible for locations in the northern hemisphere)

Materials

compass
pencil
Super Simple Science Experiments Laboratory Notebook

Experiment

❶ On a clear dark night go outside, and using your compass, find "north." North is where the needle of your compass points. Orient your body so that you are facing north.

❷ Look up at the sky and find a lone star with no bright stars nearby. This is Polaris, also called the North Star or the Pole Star. For the continental United States, it can be found about 45 degrees from the horizon.

❸ Observe this star for one week. Does the star move?

Results and Conclusions

Polaris is important because it is almost exactly above the North Pole of the Earth. You can think of the North Pole extending up to Polaris, thus making a pole in the sky. Because Polaris is at the end of this pole in the sky, it is the only star that does not appear to move. In other words, the Earth rotates around this pole, and so the North Star remains practically in the same place in the sky each night.

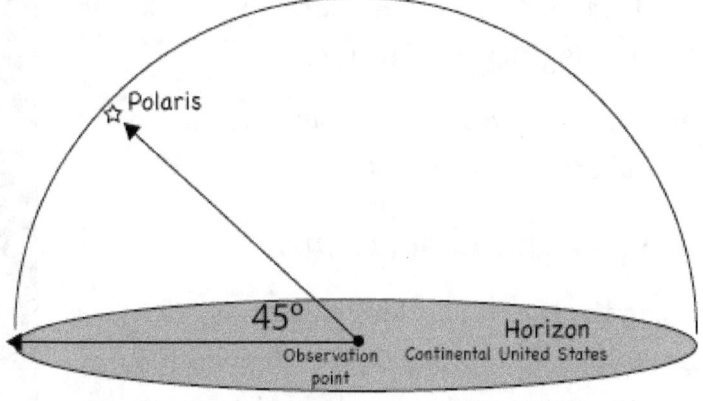

2. Finding the Big Dipper

observation

Objective

To locate the star asterism called the Big Dipper. (Best observed in the northern hemisphere during the spring)

(Note: an asterism is a group of stars that is not officially considered a constellation.)

Materials

compass
pencil
Super Simple Science Experiment Laboratory Notebook

Experiment

❶ On a clear dark night go outside and find the North Star, also called Polaris or the Pole Star. If you don't know where the North Star is, you can use a compass to find "north." Once you find north, locate the lone star that has no other nearby stars. This is the North Star.

❷ The Big Dipper is a star asterism shaped like a dipping spoon and near the North Star.

❸ To find the Big Dipper, using your imagination draw a line from the North Star to intersect two nearby stars. These two stars are the edge of the dipping spoon of the Big Dipper.

❹ Find the five other stars that make up the Big Dipper asterism.

❺ Record your observations in your laboratory notebook

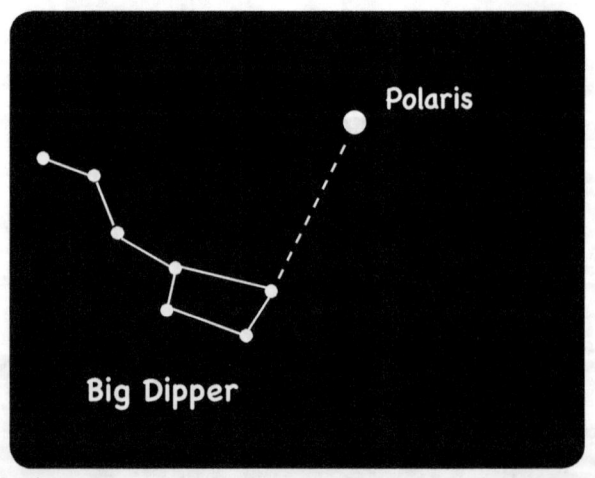

Results and Conclusions

Using the North Star as a guide, it is relatively easy to find the Big Dipper. The position of the Big Dipper will vary depending on where you live and on the time of year.

3. Finding the Little Dipper

observation

Objective

To locate the Little Dipper constellation. (Visible for locations in the northern hemisphere and best observed during the spring.)

Materials

compass
pencil
Super Simple Science Experiments Laboratory Notebook

Experiment

❶ On a clear dark night go outside and find the North Star, also called Polaris or the Pole Star. If you don't know where the North Star is, you can use a compass to find "north." Once you find north, locate the lone star that has no other nearby stars. This is Polaris, the North Star.

❷ Polaris forms the end of the handle for the Little Dipper. Most of the stars in the Little Dipper are faint, so this can be a difficult constellation to find.

❸ To find the Little Dipper, in your imagination draw a line from Polaris, the North Star, through two nearby stars that form the handle. Locate the four other nearby stars that form the dipper.

❹ Record your observations in your laboratory notebook.

Results and Conclusions

The Little Dipper is much harder to find than the Big Dipper, but it contains the North Star, Polaris.

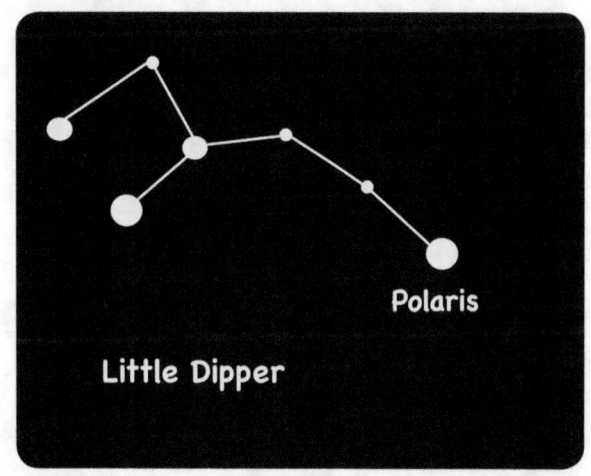

4. Finding The Dragon

observation

Objective

To locate the constellation called The Dragon. (Visible for locations in the northern hemisphere and best observed during the spring.)

Materials

compass
pencil
Super Simple Science Experiments Laboratory Notebook

Experiment

❶ Locate the North Star (Polaris), the asterism called The Big Dipper, and The Little Dipper constellation.

❷ The Dragon is a large constellation between The Big Dipper and The Little Dipper.

❸ To find The Dragon, locate four stars between the handle of the Big Dipper and the bowl of the Little Dipper. These four stars form the tail. Draw an imaginary line from the tail through a curved set of stars to the "feet." The feet of The Dragon are made of four stars that form a box. From the feet draw an imaginary line to the "head."

❹ Record your observations in your laboratory notebook.

Results and Conclusions

The Dragon is a large constellation with fairly faint stars. It can be difficult to see. The head of The Dragon is easily confused with the bowl of The Big Dipper. However, on a clear night far away from city lights, The Dragon is fun to find.

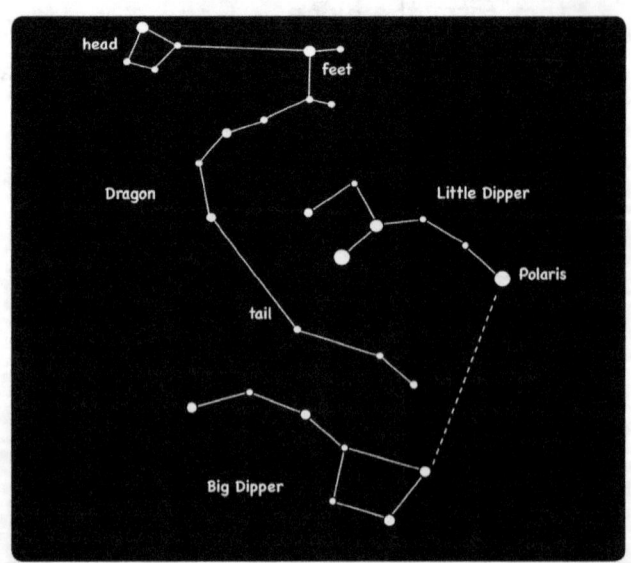

5. Finding Cassiopeia

observation

Objective

To locate the constellation Cassiopeia (Visible for locations in the northern hemisphere and best observed during the spring)

Materials

compass
pencil
Super Simple Science Experiments Laboratory Notebook

Experiment

1. Locate the North Star (Polaris).
2. Cassiopeia is easy to remember because it is in the shape of a "W" or "M" depending upon its position.
3. To find Cassiopeia, draw a line from Polaris to a star 120 degrees on the opposite side of The Big Dipper.
4. Record your observations in your laboratory notebook.

Results and Conclusions

Cassiopeia and The Big Dipper are roughly opposite each other and rotate around Polaris during the course of the night.

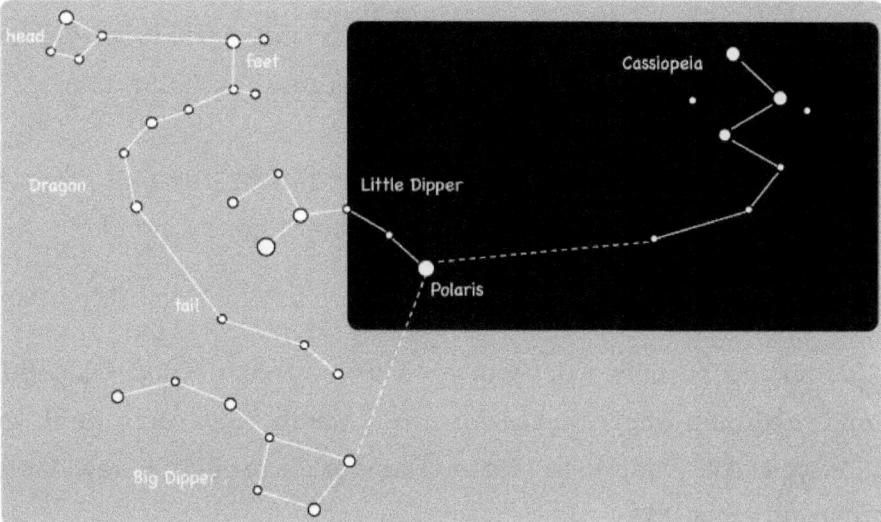

6. Moon Phase Calendar

observation

Objective

To create a Moon Phase Calendar by observing the phases of the Moon for one month.

Materials

pencil
laboratory notebook
monthly calendar template (found at the back of this book)
Super Simple Science Experiments Laboratory Notebook

Experiment

❶ Go outside on a clear night and locate the Moon.
❷ In your notebook draw the Moon as you actually see it. Record any interesting details you observe. Record the time and note the date.
❸ For the next 30 days observe the Moon at the same time each night. Record your observations in your notebook.
❹ When you have recorded the Moon for 30 days, use the monthly calendar template to create a Moon Phase Calendar. To create the Moon Phase Calendar, do the following:
  On the template, go to Day 1 and mark this day "New Moon."
  Count 7 days into your calendar and mark this day "First Quarter."
  Count 14 days into your calendar and mark this day "Full Moon."
  Count 22 days into your calendar and mark this day "Last Quarter."
❺ Look at your moon drawings and redraw the Full Moon on the Full Moon day on your Moon Phase Calendar.
❻ Fill in the moon drawings before and after the Full Moon. Record the date for each moon. You now have a complete Moon Phase Calendar.

Results and Conclusions

You should be able to create a complete Moon Phase Calendar by observing the moon, and recording what you see for a full month. How accurate is your Moon Phase Calendar? Can you use your Moon Phase Calendar to predict a new Moon Phase Calendar for the following month?

7. Solar eclipse

observation
model building

Objective

To create a model of a solar eclipse.

Materials

    pencil
    Super Simple Science Experiments Laboratory Notebook
    ping-pong ball
    softball
    flashlight

Experiment

1. Place the softball on a flat surface in a dimly lit room.
2. Place the ping-pong ball a few feet from the softball.
3. Place the flashlight facing the ping-pong ball and opposite the softball.

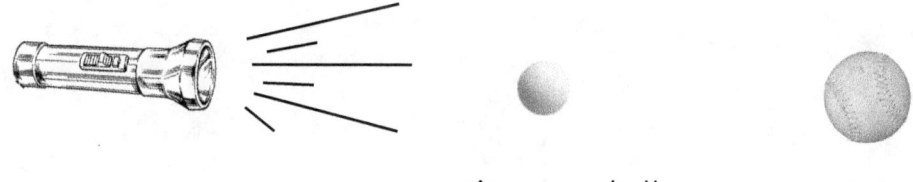

ping-pong ball   softball

4. The ping-pong ball and softball should be directly in line with each other.
5. Using your imagination, pretend you are very small and can stand on the light side of the softball. From this position, observe the ping-pong ball. (Do not look directly into the light of the flashlight.)
6. Record your observations in your laboratory notebook.

Results and Conclusions

A solar eclipse occurs when the Moon passes between the Sun and the Earth. In your model, the ping-pong ball represents the Moon, the softball represents the Earth, and the flashlight represents the Sun. Because the Moon is between the Sun and the Earth, you cannot see the Sun itself. Special glasses must be used when viewing a solar eclipse to protect the eyes from the bright sunlight that is not covered by the Moon.

8. Lunar eclipse

observation
model building

Objective

To create a model of a lunar eclipse.

Materials

pencil
Super Simple Science Experiments Laboratory Notebook
softball
ping-pong ball
flashlight

Experiment

1. Place the softball on a flat surface in a dimly lit room.
2. Place the ping-pong ball a few feet from the softball.
3. Place the flashlight facing the softball and opposite the ping-pong ball.

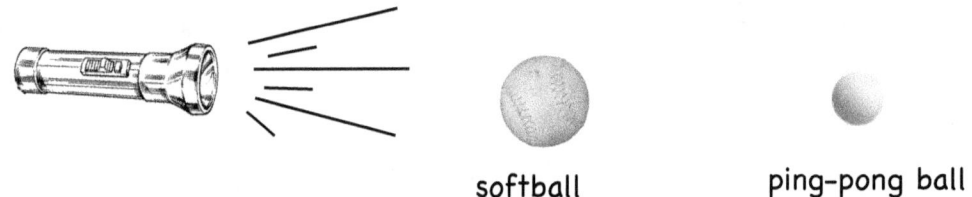

softball   ping-pong ball

4. The ping-pong ball and the softball should be directly in line with each other.
5. Using your imagination, pretend you are very small and can stand on the dark side of the softball. From this position, observe the ping-pong ball.
6. Record your observations in your laboratory notebook.
7. Experiment by changing the distance between the softball and the ping-pong ball. Does this change your observation?

Results and Conclusions

A lunar eclipse occurs when the Earth passes between the Sun and the Moon. In your model, the ping-pong ball represents the Moon, the softball represents the Earth, and the flashlight represents the Sun.

9. Ocean Waves and the Moon

observation
model building

Objective

To observe how tidal waves are formed by the rotation of the Earth and the gravitational pull of the Moon.

Materials

pencil
Super Simple Science Experiments Laboratory Notebook
balloon filled with water

Experiment

❶ Hold the water balloon in your hands. Gently rock the water balloon, and feel the water move.
❷ Holding the water balloon in your hands, spin around three times, and then quickly stop. (Don't get dizzy!)
❸ When you stop, feel the water in the water balloon with your hands. Observe how the water moves.
❹ Next take the balloon and gently pull on each end. Rock the balloon back and forth and feel the water with your hands.
❺ Keeping the balloon elongated (by gently pulling on the ends), spin around three times, and quickly stop as in Step ❷.
❻ Feel the water in the balloon with your hands as you stop.
❼ Record your observations in your laboratory notebook.

Results and Conclusions

Waves are the result of both the Moon's gravitational pull and the Earth's rotation. When the Earth spins, the water in the oceans spins too. When you stop spinning in Step ❷, you can feel the water inside the balloon creating a wave.

The Moon gently pulls on the Earth, distorting the distribution of ocean water. This distortion creates two high tides and two low tides. When you spin with the water balloon elongated (Step ❺), there is more water in the ends of the balloon than in the center, simulating the Moon's gravitational pull.

10. Building a Horizontal Sundial

using science tools

Objective

To build a horizontal sundial and use the Sun as a method for keeping time.

Materials

- cardboard (big enough to make a circle 6 inches in diameter)
- scissors
- popsicle sticks
- tape
- sundial diagram (found at the back of this book)
- protractor
- compass
- Super Simple Science Experiments Laboratory Notebook

Experiment

1. Cut the cardboard into a circle 6 inches in diameter.
2. Cut out the sundial diagram. Glue it to the cardboard circle, centering it.
3. Find the latitude of your location using an internet reference or atlas.
4. The popsicle sticks will be used to make the hand (called the gnomon) of your sundial. Place one popsicle stick along the noon hour meridian (marked by a thick black line), and tape the end of it to the middle of the sundial diagram. Place the flat edge of your protractor on the cardboard next to the popsicle stick. Keeping the end at the center of the circle fastened, lift the outside end of the popsicle stick. Using the protractor, make an angle between the stick and the paper that matches the latitude of your location. Use a second popsicle stick to support the angled popsicle stick, and tape it in place.
5. Using the compass, find due north. Point the gnomon of your sundial so it is facing due north.
6. Read the time by looking at the shadow the gnomon casts on the sundial.

Results and Conclusions

A horizontal sundial can be used to measure time. Horizontal sundials are not very accurate for locations near the equator (25 degrees latitude or less). As the Earth rotates, the angle of the Sun will change, and the shadow cast by the gnomon will shift on the face of the sundial.

11. Model of the Moon

model building

Objective

To visualize the Moon by building a model of the Moon showing different features.

Materials

styrofoam ball 1 inch in diameter
white paint
gray paint
small knife
Super Simple Science Experiments
  Laboratory Notebook

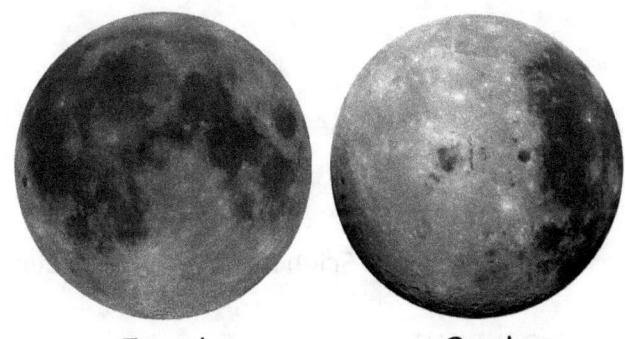
Front   Back

Experiment

1. Look closely at the front and back images of the Moon.
2. Pick one side of the styrofoam ball to represent the front face. The other side will represent the back face.
3. The dark patches on the Moon are shallow craters. Using the small knife, cut out shallow sections that looks similar to the craters in the photo.
4. Turn the ball around and repeat for the back face.
5. Using the gray and white paint, color the styrofoam ball. The shallow cut out sections can be filled in with gray paint to make them darker.

Results and Conclusions

Model building is an important part of scientific investigation. Determine how close your model looks to the real image of the Moon. What problems did you encounter with the model? How could you improve your model by using other materials or more detailed images?

12. Model of the Sun

model building

Objective

To use model building as way to learn about the Sun.

Materials

styrofoam ball 12 inches in diameter
yellow paint
red paint
Super Simple Science Experiments Laboratory Notebook

Experiment

❶ Look closely at the image of the Sun.

❷ The Sun is made of helium and hydrogen gas. The features of the Sun change over time, but there are consistently swirls, light spots, and dark spots.

❸ Paint the styrofoam ball with the yellow paint. Let the ball dry.

❹ Using the red paint, make swirls on top of the yellow paint until you are satisfied your model looks similar to the image.

❺ Let the paint dry.

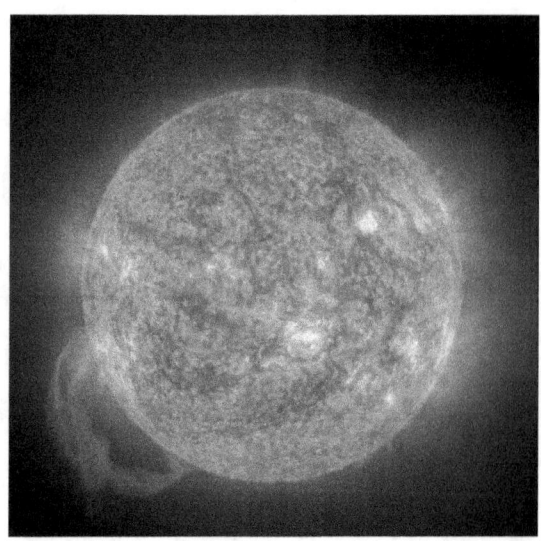

Results and Conclusions

Model building is an important part of scientific investigation. Determine how close your model looks to the real image of the Sun. What problems did you encounter with the model? How could you improve your model by using other materials or more detailed images?

13. Model of Mars

model building

Objective

To use model building as a way to learn about Mars.

Materials

- styrofoam ball 2 inches in diameter
- small knife
- gray paint
- red paint
- Super Simple Science Experiments Laboratory Notebook

Experiment

1. Look closely at the image of Mars.
2. Mars is a terrestrial planet, like Earth. Mars has an abundance of iron oxide which gives it a red color. It has volcanoes, valleys, deserts and polar ice caps, like Earth.
3. Look at the image and create a surface that looks similar.
4. Using the red and gray paint, color the surface similar to that in the image.
5. Let the paint dry.

Results and Conclusions

Model building is an important part of scientific investigation. Determine how close your model looks to the real image of Mars. What problems did you encounter with the model? How could you improve your model by using other materials or more detailed images?

14. Model of Venus

model building

Objective

To use model building as a way to learn about Venus.

Materials

styrofoam ball 2 inches in diameter
gray paint
red paint
white paint
Super Simple Science Experiments
  Laboratory Notebook

Experiment

❶ Look closely at the image of Venus.

❷ Venus is a terrestrial planet, slightly smaller than Earth. Venus has a thick, toxic, sulfuric acid atmosphere, and no light reaches the surface.

❸ Looking at the image and using the red, gray, and white paint, create a surface that looks similar.

❹ Let the paint dry.

Results and Conclusions

Model building is an important part of scientific investigation. Determine how close your model looks to the real image of Venus. What problems did you encounter with the model? How could you improve your model by using other materials or more detailed images?

15. Model of Jupiter

model building

Objective

To use model building as a way to learn about Jupiter.

Materials

styrofoam ball 8 inches in diameter
orange paint
red paint
white paint
Super Simple Science Experiments
   Laboratory Notebook

Experiment

1. Look closely at the image of Jupiter.

2. Jupiter is a Jovian planet, much larger than Earth and made of gas. Jupiter has distinct light and dark regions along the circumference. The lighter regions are called "zones" and the darker regions are called "bands." Jupiter also has a signature "Great Red Spot" in the lower hemisphere.

Great Red Spot

3. Use orange, red, and white paint to create a surface that looks similar to the image of Jupiter.

4. Let the paint dry.

Results and Conclusions

Model building is an important part of scientific investigation. Determine how close your model looks to the real image of Jupiter. What problems did you encounter with the model? How could you improve your model by using other materials or more detailed images?

16. Model of Saturn

model building

Objective

To use model building as a way to learn about Saturn.

Materials

styrofoam ball 6 inches in diameter
blue paint
green paint
white paint
toothpicks
construction paper
glue
Super Simple Science Experiments Laboratory Notebook

Experiment

❶ Look closely at the image of Saturn.
❷ Saturn is a Jovian planet, smaller than Jupiter and larger than Earth. Like Jupiter, Saturn is made of gas. Saturn has distinctive "rings" circling its midsection.
❸ Take the construction paper, and cut out several paper rings larger than the diameter of the styrofoam ball. Using paint, color them blue, green and white.
❹ Use blue, green, and white paint to create a surface similar to the surface shown in the image.
❺ Take the toothpicks and attach them to the middle of the styrofoam ball so they point out horizontally in all directions. You will use these to attach the rings.
❻ Attach the construction paper rings by laying them on top of the toothpicks and securing them with glue.

Results and Conclusions

Model building is an important part of scientific investigation. Determine how close your model looks to the real image of Saturn. What problems did you encounter with the model? How could you improve your model by using other materials or more detailed images?

17. Model of Neptune and Uranus

model building

Objective

To use model building as a way to learn about Neptune and Uranus.

Materials

- 2 styrofoam balls 4 inches in diameter
- blue paint
- green paint
- white paint
- Super Simple Science Experiments Laboratory Notebook

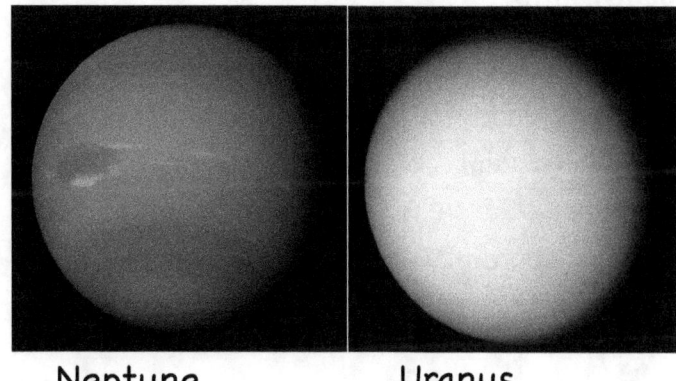

Neptune   Uranus

Experiment

1. Look closely at the images of Neptune and Uranus.
2. Both Neptune and Uranus are Jovian planets and are made of gas. They are similar in size (Neptune is slightly larger) and are blue in appearance. Neither have significant surface features.
3. Use blue, green, and white paint to create a surface similar to the surface shown in the image.
4. Let the paint dry.

Results and Conclusions

Model building is an important part of scientific investigation. Determine how close your model looks to the real images of Neptune and Uranus. What problems did you encounter with the model? How could you improve your model by using other materials or more detailed images?

18. Model of Mercury

model building

Objective

To use model building as a way to learn about Mercury.

Materials

styrofoam ball 1.5 inches in diameter
gray paint
white paint
Super Simple Science Experiments
  Laboratory Notebook

Experiment

❶ Look closely at the image of Mercury.

❷ Mercury is the smallest planet in the solar system. Mercury is a terrestrial planet and orbits closest to the Sun. The surface has craters, like the Moon, with little or no other visible features.

❸ Use the gray and white paint to create a surface similar to the surface shown in the image.

❹ Let the paint dry.

Results and Conclusions

Model building is an important part of scientific investigation. Determine how close your model looks to the real image of Mercury. What problems did you encounter with the model? How could you improve your model by using other materials or more detailed images?

19. Earth

model building

Objective

To use model building as a way to learn about Earth.

Materials

a styrofoam ball 2 inches in diameter
blue paint
green paint
white paint
brown paint
Super Simple Science Experiments Laboratory Notebook

Experiment

❶ Look closely at the image of Earth.

❷ Earth is the third planet that orbits the Sun. Earth is the only planet known to support life. It has a multicolored surface that is brown, green, blue, and white.

❸ Use the green, blue, brown and white paint to create a surface similar to the surface shown in the image.

❹ Let the paint dry.

Results and Conclusions

Model building is an important part of scientific investigation. Determine how close your model looks to the real image of Earth. What problems did you encounter with the model? How could you improve your model by using other materials or more detailed images?

20. The Solar System

model building

Objective

To use model building as a way to learn about the solar system.

Materials

the planet, Moon, and Sun models made in Experiments 11-19
large cardboard surface
glue
marker
Super Simple Science Experiments Laboratory Notebook

Experiment

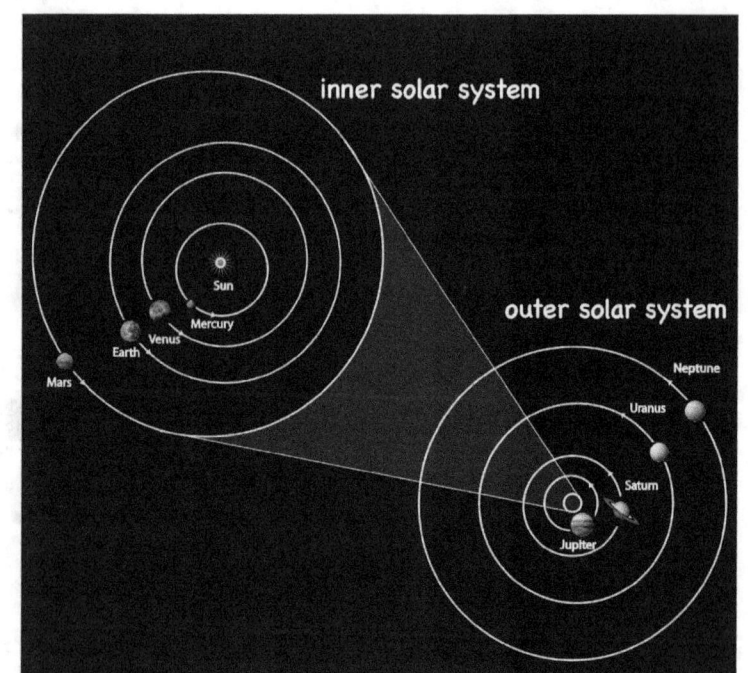

❶ Place the Sun model in the center of the cardboard surface. Glue in place.

❷ Look at the orbital path for each of the planets. Allowing several inches between orbits, draw an orbit for each planet around the Sun.

❸ Place each of the planets on its respective orbit. Glue in place.

Results and Conclusions

Model building is an important part of scientific investigation. Determine how close your model looks to the orbital paths of the planets in the solar system. What problems did you encounter with the model? How does the size of each planet and its orbit make it challenging to build a model of the entire solar system with accurate scale?

21. Find the Center of the Milky Way Galaxy

science tools

Objective

To use a computer program to locate the center of the Milky Way Galaxy.

Materials

pen
computer and internet service
Google Earth
Super Simple Science Experiments Laboratory Notebook

Experiment

❶ Set up Google Earth on your computer.
   ① Go to http://earth.google.com and click "Download Google Earth."
   ② Click "Agree and Download."
   ③ Once the file has been downloaded, install the program.
   ④ Open the Google Earth program on your computer.
   ⑤ Set up Google Earth in Sky Mode.
   ⑥ At the top, click "View" and then click "Switch to Sky."
   ⑦ On the left-hand side of the window, you should see "Layers."
   ⑧ Uncheck every item, except "Imagery" and "Backyard Astronomy."
   ⑨ Click the arrow next to "Backyard Astronomy."
   ⑩ Uncheck every item except "Constellations."

❷ Spend a few minutes becoming familiar with the Google Earth program. Toggle the icon for Earth Mode and Sky Mode.

❸ While in Sky Mode type in the phrase "Galactic Center" in the "Search the Sky" option. Zoom in and out to explore the galactic center.

Results and Conclusions

Google Earth is a useful computer tool for studying the Earth and the sky.

Sundial diagram for Experiment 10

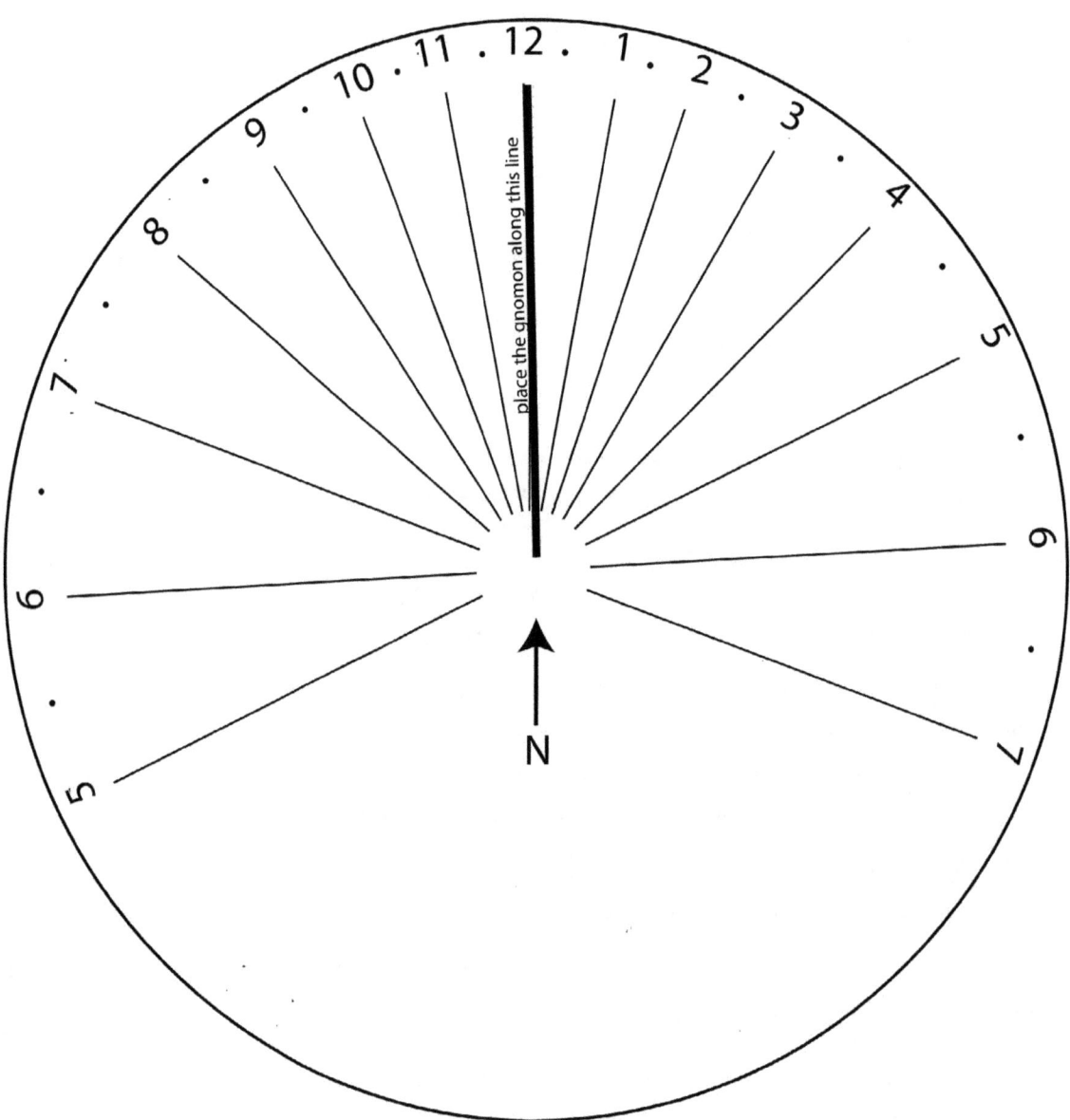

Using scissors, cut out this sundial diagram. See complete directions in Experiment 10 of this book.

Astronomy

Day 1	Day 2	Day 3	Day 4	Day 5	Day 6	Day 7
Day 8	Day 9	Day 10	Day 11	Day 12	Day 13	Day 14
Day 15	Day 16	Day 17	Day 18	Day 19	Day 20	Day 21
Day 22	Day 23	Day 24	Day 25	Day 26	Day 27	Day 28
Day 29	Day 30	Day 31	Notes			

www.ingramcontent.com/pod-product-compliance
Lightning Source LLC
LaVergne TN
LVHW082059090426
835512LV00038B/2599